Advance Study Guides of the Scientific Age

Miguel A. Sanchez-Rey

Table of Contents

Preface

4-7

Advance Quantum Geometrodynamics Of A Much Simples Quantum Foam

9-10

No-Boundary Proposal in Quantum Space-Time

12-13

A Modern Dystopia in the Scientific Age

15-17

The Racist Society

19-24

Utopian Science and Anarcho-Ecology

26-29

CompContr, LF, and Incalculability [First and Foremost]

31-32

Prospectus of the War-Machine

34-37

A Phenomenological Comparison

39

The Messy Physics Map

41-43

Preface

Almost 8 years has passed since my first publication "On The Principles of Mathematical Linguistics," in the Web Journal of Formal, Computational, and Cognitive Linguistics. A three-year effort to understand the rudimentary building block of the mind/brain and to complete the Biolinguistics Framework. Leading to the resolution of DARPA's [Defense Advance Research Projects Agency] number one mathematical grand challenge to find a functional model of the human brain. A painstaking process of careful mathematical research and problem solving. That eventually led to the abandonment of the language gene and the heralding of advance artificial intelligence in 2017.

A six-month effort in 2011 was spent studying higher-energy physics and string theory in hopes of solving the grand unified theory and the nature of strings. Leading to the publication at Vixra.org, "The Logical Structure of Space-Time." Which led to the discovery of a variant [of stringy] (initially thought of as the confirmation of the existence of strings and the metaphysical conjecture of a universal law of nature. That sought to put an end to the circular logic inherent in the dilemma of where string theory came from. Thus explaining the nature of both space-time and prespace-time physics).

A two-year hiatus later led to, "The Foundations of Quantum Field and It's Particulates" which was a rewarding effort to solve the Yang-Mills and Mass Gap Conjecture: find an axiomatic proof of Yang-Mills gauge theory and to determine the Mass Gap. Two years later the chaos of the 2015 Nobel Prize led to the hypothesis of the horrific outcome. Leading to the founding of The Physicalist Program [PHPR] as the resolution to a foreseeable catastrophic scenario in the Scientific Age in the form of a task.

In conjunction the completion of *Isaac's Laws:* an experimental work of literature (a four-year ordeal), aim to determine (through a literary experimental process) what went wrong with Post-Modernity after the chaos at CERN's (that led to the discovery of the existence of the Higgs-Boson particle. Eventually revealing that the Standard Model was inconsistent with the string models that were used to corroborate the existence of supersymmetric particles. Casting SUSY into doubt. With the Standard Model being more consistent with experiment than with SUSY), Large Hadron Collider. And at the time, the possibility that the next epoch, -- the Scientific Age, will be an age when high-expectations aren't realized.

Leading to the seven resolutions of the Nazi state and various other resolutions quintessential to the original project of Post-Modernism. A project to find a transcendental philosophy that will surpass the sciences and lead to a paradigm shift in the very psyche of human phenomenology.

The original intent of PHPR was to prevent the Scientific Age. Yet the LIGOS experiment -- that finally found evidence of the existence of gravitational waves, proved to be a catastrophe that led to a maniacal world-wide shock wave in the very consciousness of the historical consciousness. Causing subtle (yet hauntingly noticeable) havoc at a world-wide scale. Ushering the Scientific Age in which the planet was driven over the edge by flaw decision-making. Flaw decision-making that can be seen as the maturation of a Cloud-Atlas mosaic (or the sociopathic norm) -- exacerbated by the introduction of the internet and lower academic standards at the dawn of the Windows XP revolution.

Causing instability to the scientific process and eventually plunging the scientific process into complete disarray -- with the rising threat of the sociopathic norm and a pathological mass-movement, simultaneously both the decline of the religious state and Earth's biosphere (as a result of global warming and mineral depletion).

PHPR became the imperative option to contain a planet in praecox. To snap the general public out of collective psychosis through a joint effort (with the national laboratories) to periodically shut the general public from the scientific process. To put into place three tasks of PHPR -- two tentatively, that aim to achieve three monumental resolutions: mineral depletion, an extraterrestrial conflict, and the healthy transition into an internationalist model at a interplanetary scale.

These three tasks set the bases for a world-changing program that sets precedent for the founding of the planetary superstate, the quiet dismantlization of the religious state, and the ushering of Anarcho-syndicalism. An arduous process that requires a wildly-strong predisposition.

Eventually the discovery of metaspace, the realization of Utopian Science, the application of the Omega-Kardeshev scale, and the logical conclusion of the nightmarish outcome. All my works are a painful and difficult effort to implement crisis control in decision-making and leadership. Which gave way to the founding of The Academy of Advance and the Technological Sciences [AASTS]. The AASTS – the logical conclusion of *Utopian Science*, is a many century long effort to control and mitigate the threat of unstable decision-making which originally cause mayhem to the scientific process. A promising academy of a prideful regret. Stated more carefully in *Thoughts of the Scientific State* and *An Autobiographical Study in the Scientific Age*.

Other discoveries and explanations yielded the interplay and logical intuitionism – especially natural Platonism. Anarcho-syndicalism as being a plausible outcome of the religious naturalist movement. And the beginning of Advance Physics as the resolution to a crisis in the theoretical sciences.

All my current works are templates design for others to read, delve into, apply and enjoy. To increase problem-solving, pattern recognition and comprehension. Leaving others, the responsibility of proper, precise and honest examination (most especially The Leading Professors of the AASTS).

To be applied (carefully and thoughtfully) in an effort to understand what to search for, and how to cope with, both the dawn and the arrival of the Advance Age. And to make way for further psychological studies and scientific advances in PHPR that will pave the way to a radiant future for humanity. Never be too nostalgic of the past. Instead be more open minded of a brighter hereafter.

- The Leading Professor Miguel Angel Sanchez-Rey [*The Grandmaster, The Master of Space-Time*]

Advance Quantum Geometrodynamics Of A Much Simpler Quantum Foam

The Leading Professor Miguel Angel Sanchez-Rey [*The Grandmaster*, *The Master of Space-Time*]
The Physicalist Program
The Academy of Advance Science and the Technological Sciences

Abstract

Discussing advance quantum geometrodynamics of a much simpler quantum foam.

January 25th, 2018

Advance Quantum Geometrodynamics [AQG] is a study on John Archibald Wheeler's, Kip S. Thorne's, and Charles W. Misner's textbook, *Gravitation*. The underlying premise is that, *Gravitation* is a misnomer (a bloated survey textbook on general relativity and modern gravitational cosmology).

That quantum geometrodynamics has been resolved as the manifestation of an closed superstring, call the graviton, through the super-Yang Mills gauge analog. In that way, advance quantum geometrodynamics is ascertain and establish as a much simpler and more important open problem of advance physics. A problem which aims to put advance physics on a much simpler quantum foam that becomes more and more simpler as one projects further into the advance quantum spin-supermanifold.

An aspect of quantum topology, advance quantum geometrodynamics means the possibility to manipulate gravitation and to use AQG to further knowledge on super-black holes which are black holes that carry supersymmetric properties and which are said to be black holes that not only causes space-time to breakdown but also causes matter to become more and more heated and energetic -- to the extent that matter ignites and disintegrates in a molten, violent and turbulent event horizon. AQG also aims to answer questions that will lead to a simpler quantitative theory of physics.

A simpler quantitative theory of physics that will accelerate advances in theoretical physics that will supersede all past advances in classical and modern physics. Revitalizing high-energy physics -- in such a way, that its applications will yield astronomical technological advances that coincides with PHPR [The Physicalist Program]'s, The Grand Unification Scheme. Which aims to manipulate matter at the quantum level, or at the Ad [superstring]-level, by achieving the terraformic process. With AQG it will be much easier to not only manipulate matter using The Grand Unification Scheme but also to take advantage of other forms of matter; including exotic (carrying significant implications for much of what is understood as the bases of quantum reality and prespace-time physics).

The implications of AQG are limitless and worthwhile -- to pursue significant research on the mathematical properties of advance quantum spin-supermanifold that carries heavy implications for dynamical systems, symplectic morphic-geometry and quantum topology. Giving life to the field of pure mathematics and also revitalizing the field of geometrodynamics into a more refine and open problem of advance physics that will have a lasting effect on the entire field of physics for many centuries to come.

The No-Boundary Proposal in Quantum Space-Time

The Leading Professor Miguel Angel Sanchez-Rey [*The Grandmaster*, *The Master of Space-Time*]
The Physicalist Program

Abstract
What came before the initial state that led to the inflationary early universe?

March 5th, 2018.

The Hartle and Hawking No-Boundary Proposal states that the universe has no boundary. And that any event in space-time is precluded and surpassed by an earlier event and a future event (and vice versa). That space-time is a quantum state of a vast quantum universe where the wave-function of a quantum state of a universe has larger values and that improbable universes has smaller if zero values. Giving the continuum paradox Minkowski space-time is the sum of the dot-product of two events without end. In which, imaginary time transforms Minkowski space-time into quaternion space-time that contains the cosmological wave-function of an inflationary state of a self-referential quantum universe. Where imaginary time is neither imaginary nor real. But self-reflexive -- in such a way, that space and time is without a boundary or a continuum. Implies quantum space-time is null.

A Modern Dystopia in the Scientific Age

The Leading Professor Miguel Angel Sanchez-Rey [*The Grandmaster*, *The Master of Space-Time*]

The Academy of Advance Science and the Technological Sciences

The planetary biosphere has decline in recent months as much of the world has undergone a praecox shock that has cause humanity's thought processes to become permanently graphic and intense. Causing violent antisocial behavior to become more frequent and erratic. With a looming environmental crisis and a decline in biodiversity -- due to deforestation cause by human overpopulation, an outcome of global warming, much of what is understood as the norms of Post-Modernist thought has collapse into a planet that has reach the Scientific Age.

Giving the fall-out of scientific radicalism and the counter-reaction to pure philosophy, a pathological mass-movement has rendered the expectation of Post-Modernist social, ethical, and political values incapable. That is what seemingly stands as a planet in recovery, whether or not any statistical data can show a decline in violence, is in actual quiet mayhem and volatility. Where the political and scientific process cannot reconcile ideology, of both neo-Fascism and social progressivism, with the skeptical inquiry and progressive sentimentality of the philosophical sciences.

As the planet now resides as a modernist world living in a Dystopian reality in the Scientific Age. For what seems, on a large-scale, as the beginning of peaceful existence (when magnified) is, in truth, a planet where much of what is understood and observed is not what it seems. And for which much of humanity presides in a Dystopian reality that is meager and counterproductive -- both in its inability to actualize human potential and market forces. Human potential and market forces that are in favor of natural and human rights.

For those reasons neo-liberalism, academic elitism, and establishment politics has failed to realize expectations in a planet where humanity does not achieve said expectation, rather peace and tranquility is nothing more than personal conflict and universal despair. A despairing fact that is beholden to a truthful realization of desperation and extremism in both the sciences and the political domain.

The Racist Society

The Leading Professor Miguel Angel Sanchez-Rey [*The Grandmaster*, *The Master of Space-Time*]

The Academy of Advance Science and the Technological Sciences

The planet has look endlessly in denial about the unstable historical notions that has captivated much of the civil population of the Western nation-states. Looking upon itself as vindicated scientific radicalist that flirts with much of the world into implementing their visionary hopes of a neo-fascist scientific regime. A scientific regime that yearns to incite social neo-Darwinism upon the civilized world and that what appears as, "men of science" are, in factual truth, aristocratic members of a dying statist ideology, that is desperate to maintain a nationalist spirit in order to preserve both class privilege and racial supremacy.

Volatile decision-making has engulfed the global market economy, and at both ends of the Atlantic, governments (considered to be constitutional democracies that share an interrelated lineage of Anglo-Saxon culture and Christian dogma) has turn to destructive policy-making without any hindsight.

That is nothing much has been done, and what appears as the life-world succumbing to meta-philosophical reasoning, has resulted in the planet falling victim to a false sense of euphoria. Only as a consequence has the general population succumb to psychotic literature. But has also accepted the propaganda machines of special and powerful interests. Psychotic literature and propaganda machines that unleashes upon the planet the near uncontrollable epidemic of a pathological mass-movement that believes it can rise above false decision-making. Only causing a planet, seemingly at ease, to fall from grace into a modern dystopia in the Scientific Age.

For what is the nature of this dystopia, in which, the radicalization of the sciences is so embraced by the powerful elites that they set to shield the scientific radicalist, and the pure philosophers, from honest ridicule and true contempt. For very little humanism persist in the civilized Anglo-Saxon world.

A humanism, as Corliss Lamont eloquently conceptualized to be rooted -- in not only the freedom of speech, but also in civil activism, criticism of state and religious tyranny, and the protection of the general population from self-destructive private interest that focuses too much on short-term profits.

The racist society has been embraced in open arms by both neo-liberalism and neo-conservatism. And upon both ends, the racist society has welcome the modernist dystopian reality of the Scientific Age. As a precursor to a global dynasty class of scientists and philosophers whom yearn for a pure philosophical and a pure anarchistic ideology of class, wealth, and privilege to incite opposition against the Scientific Age. So that they may take control of decision-making, to further their fame and fortune, on spurious and dubious grounds. That is logo-centrism has been taking to the extreme. Yet racism and bigotry has become the bedrock of the sciences.

There can be no telling what anti-Chicanoism, anti-Semitism, anti-African, and/or anti-Asian sentiments may mean for much of the Anglo-Saxon world. And what path the sciences will follow, if it is to avoid racism and class privilege. Which has engulfed the learn institutions and scientific academies of the classical world (whom idealized themselves as imperial exemplification of colonialism and barbarism).

That is the United Kingdom has become the *epitome of the racist society*. And only upon learning what the nature of this racism is and what it's implications are to race relations on a planetary scale, it only dawns, upon many activists and outspoken dissidents, that a dialectical contradiction has made normalcy cruelty and racial injustice a righteous act.

As the sciences has become a radicalize branch of a rising pathological mass-movement that has invaded much of the civilized world.

Plunging the scientific process, not only into disarray, through cry havoc -- of the pillage and plunder of the masses, but also quiet mayhem (by the deception and conceit of individual self-interests). Bringing about a modern dystopia in which there is no end in sight to the false euphoria and racial bigotry.

Utopian Science and Anarcho-Ecology

The Leading Professor Miguel Angel Sanchez-Rey [*The Grandmaster*, *The Master of Space-Time*]

The Academy of Advance Science and the Technological Sciences

Utopian science is an empirical analysis of political axiomatics. Political axiomatics that is known to be true according to what constitutes a decent or indecent society in a utopia or dystopia; both respectively. As a careful empirical analysis of a modern dystopia shows that racial supremacy results in a modern dystopia in the Scientific Age.

Since the social ecology of the Scientific Age is an immeasurable ecology, Anarcho-syndicalism (being the inevitable outcome) instead helps to see social ecology in the Scientific Age as an anarcho-ecology of individual co-existence with the environment. And that individual action to protect the environment -- as an attribute of the federation of democratic industries, leads to a healthy predisposition to see social organization as harmonious with the environment. Rather than as a hierarchal obligation to political authoritarianism of the religious state.

Since the religious state is a permanently declining power-structure, and that the super-state is the resolution to individual action and the sacredness of the state, the dismantling of the religious state makes anarcho-ecology an important function of the internationalist model and the environmental democratic outlook of individual autonomy.

That the well-being of individual existence, and the necessity for co-existence (base on a harmonious power-play), makes the flourishing planetary biosphere a product of individual reorganization of social structure and the preservation of justifiable political norms. Quintessential to two political axiomatic, i.e., the democratic ownership of the environment and the rights of future generations, as a precursor to securing long-term ecological equanimity with the federation of industrial democratic economies.

Where maintaining the environment, and by applying social engineering to model an ideal scientific utopia (on the grounds of political axiomatics that is understood to preserve the biosphere), is an important construct of long-term effective decision-making. Which stresses a wild strength that is indebted to will-to-power -- of Friedrich Nietzsche, and independent co-existence of workers' councils and trade groups (of Mikhail Bakunin), that acknowledge environmental self-control as an important element of anarcho-ecology.

Whereas an immeasurable ecology means the visual anticipation of the social ecology of the Scientific Age is too early, or an unknown in the far distance, the anarcho-ecology of the Scientific Age better represents the social organization of an anarcho-civilization. Disciplined as individual action to protect the biosphere for future generations while uphold harmony and distributing well-being as an invaluable right.

For there is no other alternative to Anarcho-syndicalism, and for that reason anarchism politics -- in a biosphere that is maintained through individual action of trade workers and workers' councils, is an inevitable byproduct of the inevitable outcome of political axiomatic of the federation of democratic industries and anarcho-ecology of the Advance Age. Empirical analysis guided by free-markets and a common democratic lineage that can preserve the workers control of the means of production -- through democratic means, the attachment of self-embodiment, and the right to self-manage and self-actualize (on equal outcome) which is obligatory to self-ownership, individual autonomy, and scientific utopianism.

CompContr, LF, and Incalculability [First and Foremost]

The Leading Professor Miguel Angel Sanchez-Rey [*The Grandmaster, The Master of Space-Time*]
The Physicalist Program

Abstract

A no-brainer.

February 11th, 2018

Gauge theory is how modern physics deals with atypical redundancy. That is one imposes a gauge connection on a Lagrangian equation to achieve gauge invariance. For equations of motion rely on gauges to achieve gauge equations that shows symmetry -- or gauge symmetry. For even so (using integration) gauge equations achieve mathematical favorability. The information they attain leads to achieving more and more precision in high-energy physics. Allowing the quantification and modeling of physical phenomenon (quark confinement, string interactions, and spin-foams) to integrate with each other in such a way that gauge models can be constructed that yields information about particle behavior, and their composition, as one gets closer to the Planck energy scale. But even then gauge theory is without its drawbacks, for gauge symmetry is a valuable commodity but a brain drain that eventually crash the field of theoretical physics into computational disarray. With integrable functions (like the Lebesgue measure, σ-measure or the 10-dimensional S-matrix for super-string interactions) calculations become more and more burdensome. Such burdensome implies further strain at the expense of computational power and efficiency.

So the solution is to introduce logical form [LF] and computational control [CompContr] to simplify the field of physics to achieve parsimony, elegance, and eloquence. Enabling more computational power while achieving more and more precision in such a way that complex physical equations and long calculations is no more. Hence one achieves the birth of Advance Physics. As modern physics has become a marvel of classical physics and a short-live fashion statement in the sciences. That said, Incalculability is an updated approach to SUPREME. But an engineering approach. For those reasons there can be no denying, integrability has no qualifications to address the incalculable. An integrated approach that is well-understood while the least of one's concern.

Prospectus of the War Machine

The Leading Professor Miguel Angel Sanchez-Rey [*The Grandmaster*, *The Master of Space-Time*]

The Physicalist Program

The Academy of Advance Science and the Technological Sciences

The war machine often leads to a competing effort for duplicate war machines in an attempt to realize competing self-interests. The war machine is a short-live byproduct of a war-like society. If the idea of the war machine spreads like a meme -- across an entire galaxy, the war machine will eventually tear apart the whole galaxy.

A frequent occurrence amongst the cosmos but unnoticeable to most primitive life (since an advance civilization, driven by the war machine, could ultimately be controlling such primitive life). Such that, primitive life may be unaware that they are being sheltered from a conflict that wages war amongst competing war machines. Competing war machines that are endeavoring to outwit and destroy the war machine. A war machine that has led to a controlled civilization.

For those reasons, The Second Task (which has been establish as the resolution to an extraterrestrial conflict) is essential to long-term human longevity. To safely expand from Earth's solar habitat -- in order to extend human survival, and thwart a population, energy and mineral crisis within Earth's solar habitat.

To do so, The Physicalist Program [PHPR] top-scientists will likely aim to search out for habitable zones in an attempt to realize the internationalist model -- peaceful co-existence through peaceful First Contact. By achieving First Contact (in the form of a ritual), humanity will be imparted knowledge of what habitable zones are open to human settlement.

Choosing co-existence and non-interference, such intelligent beings will remain neutral to human affairs. At that point, humanity will be able to safely expand from Earth's solar habitat. Achieving the beginning stages of the internationalist model in space exploration.

Constructing a safe passage that paves the way, in the Scientific Age, to a healthy transition into the dawn of the Advance Age -- approximately a century before arriving at Class 2 in the Omega-

Kardeshev scale. An age of wild strength in which humanity will exist independently from other intelligent civilizations.

Traveling in and out of star gates and worm-holes -- yet Anarcho-syndicalist and peaceful in nature, but as close to a united federation of planets (under the stipulation of the internationalist model).

David Allison and Friedrich Nietzsche: A phenomenological comparison

The Leading Professor Miguel Angel Sanchez-Rey [*The Grandmaster, The Master of Space-Time*]

The Academy of Advance Science and the Technological Sciences

David Allison seems very much Nietzschean, in that by the standards of European continental philosophy he was an existential empiricist that enjoyed story-telling. For factual story-telling is at the heart of continental philosophy. Where continental philosophy is the study of Anglo-Saxon European philosophy. A philosophy that is cognizant of the French and German tradition. Starting with Rene Descartes (the rationalist) and Immanuel Kant (the rationalist-empiricist), that also includes the Anglo-Saxon tradition of the classical modern philosophy of Samuel Pufendorf, David Hume, John Locke, George Berkeley and etc. Yet open to the classical American pragmatism -- of Charles Sanders Peirce, William James and John Dewey, that is sympathetic to an experimental and pragmatic approach to epistemological and ethical inquiry. And also delving into the critical psychoanalysis of the historical materialism of the Frankfurt school, the theoretic of interpretation of the German philosopher Hans-Georg Gadamer, the deconstruction of language and the written text developed by Jacques Derrida, and the anti-Freudian and semiotic study on the development of language and cognition by the late Gilles Deleuze.

Philosophy is grounded on the logo-centrism of being. Being is known in Martin Heidegger's, "Being and Time" as the essence of the life-world. Modernist thinking of the 20th century incorporates being as an attribute of transcendental nihilism, in that human metaphysical reality is inherently existential. An existentiality that stresses application of Friedrich Nietzsche and Søren Kierkegaard. The fathers of existential philosophy and early post-modernist thought. Heralding the contemporary field of phenomenological studies that laid the bases for research in consciousness, perception and behavioral psychology. The interface of which is the study of language in the form of semiotics and pragmatics. Leading to the founding of existentialism by Jean-Paul Sartre.

For Friedrich Nietzsche dealt with transcendental nihilism by imposing the will-to-power and the eternal return, while Edward S. Casey dealt with transcendental nihilism by arguing in trash nuisances. For Robert P. Crease argued instead for a quantum moment by emphasizing metaphorical and enthusiastic reasoning cut off from effective long-term decision-making. The technoscience of the late Don Idhe was inadequate to the spirit of the age. Harvey Cormier's honest study on experimental pragmatism wasn't all that satisfactory. And even Donn Welton's phenomenological reduction seem overly Cartesian. It just didn't seem all that Nietzschean, but very much Nietzschean in story-telling. Their continental philosophers that resort to ineffectual story-telling to prove a scattered point.

For Friedrich Nietzsche was his own person and to be respected as his own person. Having laid the impetus for the abandonment of wild-being in favor of wild-strength. The heralding of rational existentialism and the interplay. And the unraveling of logical intuitionism (owing its inspiration to the radical empiricism of the Vienna Circle) at the beginning of the Scientific Age. For the theatre of the absurd lies in psychotic literature and the herd mentality -- and for that reason, Friedrich Nietzsche dares not get involved in a theatrical absurdity.

The Messy Physics Map

The Leading Professor Miguel Angel Sanchez-Rey [*The Grandmaster, The Master of Space-Time*]
The Physicalist Program
The Academy of Advance Science and the Technological Sciences

Abstract

The Langlands Program is a messy physics map.

March 31st, 2018

The Langlands Program was developed by Robert Langlands as a bridge between number theory and geometry -- especially differential geometry, algebraic geometry and geometric topology, that seeks unification of Galois groups to automorphic forms and algebraic groups to representation theory. Where Galois groups are specified groups with a field extension, automorphic forms are well-behaved functions from a group topology to complex forms -- invariant, and representation theory expresses abstract groups in terms of linear transformations of vector spaces.

One of the most beautiful feature of The Physicalist Program [PHPR] is that numbers (in metaspace) are no longer measurable quantifiable objects that carry single values, but through the TrH theorem are super-objects that are constantly changing and look a lot like the set of $\{\Phi, 1, 2\}$ in K-theoretics. Where Φ is the null set.

With number space of the L - parameter, measure theory becomes an operation of computational form call CompContr [computational control]. CompContr seeks to limit computation -- along with logical form [LF], in order to obviate a brain drain in gauge theory. So that calculations become more and more parsimonious, elegant, eloquent and simple.

If the Langlands Program were to be the standard of an attempt to seek unification of a kind of mathematical theory, and applied to PHPR's First Task, then a messy physics map becomes catastrophically more and more enunciated -- the Langlands Program is a catastrophe waiting to happen. Such that giving the measure $\varnothing$ [diagram], a breakdown in metaspace begins to take form. Whereby relating standard number theory to geometry becomes inadequate and redundant, while the physical sciences abhors any mathematical construct that doesn't emphasize conceptual categorical simplicity that obeys experimental physics. Especially experimental physics that seeks to develop a scheme (The Grand Unification Scheme) which contains distinctive physical properties.

The Grand Unification Scheme aims to manipulate matter at the Ad [superstring]-level. Where Langlands Program holds no direct experimental certainty but can only be inefficiently implied in mathematical logic. Where anything that comes close to mathematical completeness is automatically disqualified. As indicative of the failure of David Hilbert's program in the early 20th century.

Mappings not so much that achieves geometric number theory, but in which metaspace is homotopic and subject to a K-theory of higher qualitative form. Using Incalculability not only to model Ad [superstring]'s but also to achieve metacontainment. A resolution by way of an engineering approach. That is with Langlands program, metacontainment becomes unsustainable with imminent fatality.

Langlands Program (though a formal breakthrough in mathematical theory) is an excessive burden of mathematical ingenuity. An excessive burden of mathematical ingenuity that poses serious problems to PHPR's First Task -- in which, application is inessential and too primitive to be considered a viable alternative for Advance Physics. As Advance Physics is not a geometric accomplishment but a computational achievement (counter to the Langlands Program of mathematical redundancy) that sees modern mathematical theory as too dependent on group theory (especially non-abelian gauge theory) -- leading to a computational crisis in the theoretical sciences. Where its overuse gave way to disarray.

A confused state that spelled the end of modern physics. Modern physics that shows no care for classical mathematical principles. Classical mathematical principles that depends on any mathematical approach that obligates itself to simple computational form. Simple computational form tantamount in genuine experimental science. And by acknowledging this obvious factual truth, Langlands Program becomes an ineffective mathematical language that is devoid of practical engineering and ill-informed of experimental naturalism.

www.ingramcontent.com/pod-product-compliance
Lightning Source LLC
Chambersburg PA
CBHW062234220526
45471CB00009B/3478